Volcanoes

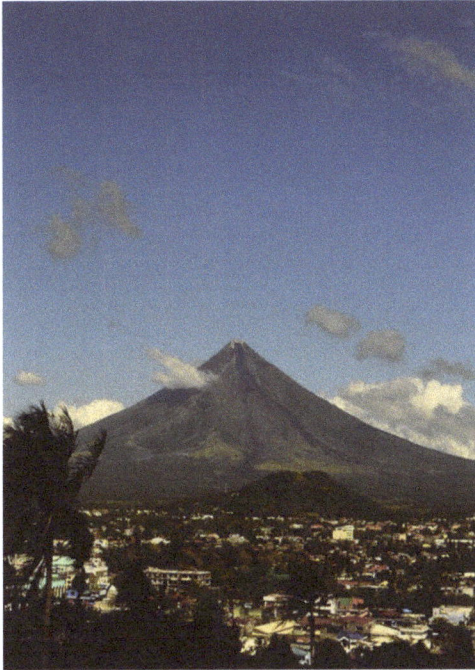

By Adelina Tibell

Library For All Ltd.

Volcanoes

First published 2021

Published by Library For All Ltd
Email: info@libraryforall.org
URL: libraryforall.org

This book was made possible by the generous support of the Education Cooperation Program and the following organisations.

Australian Aid

ChildFund Australia   PLAN INTERNATIONAL   AHP Disaster READY

Volcanoes
Tibell, Adelina
ISBN: 978-1-922550-21-7
SKU01571

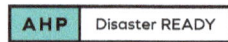

Volcanoes

What is a volcano?

A volcano is an opening in the Earth's crust that lets out gas, ash, and lava. Most volcanoes are mountains.

Crater

Ash cloud

Vents

Lava flow

Conduit

Magma chamber

A volcano

Earth has three layers: the crust (the outermost layer), the mantle (the middle) and the core (the centre).

Molten rock is called magma when it's under the Earth's crust. When it comes to the surface it's called lava.

Scientists study volcanoes to know when the next eruption will take place, so they can warn people living close to the volcano.

Why do we have volcanoes?

The Earth's crust is made of large plates. The plates move against each other, and this builds up pressure. Volcanoes erupt to let out the pressure! Volcanoes can also form in hot pots.

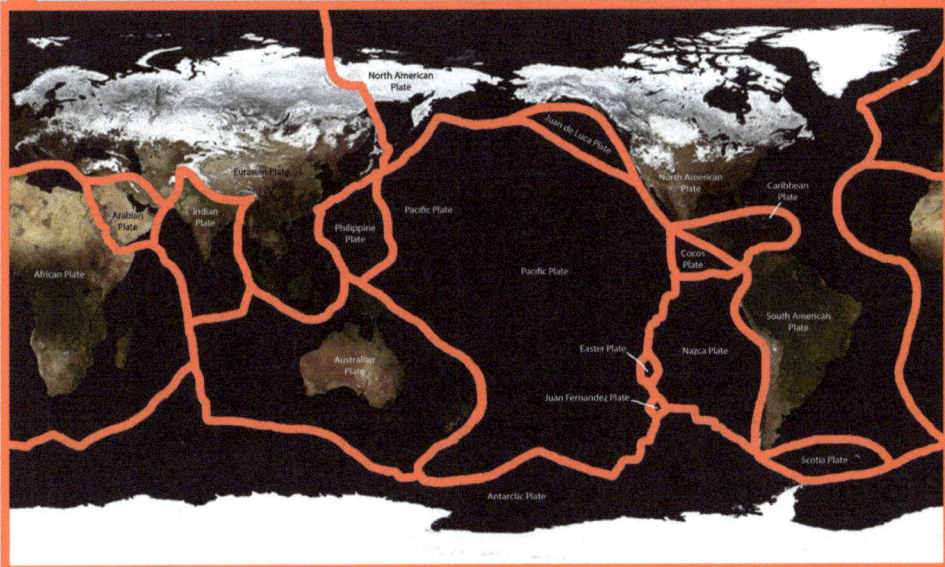

Eurasian Plate
Arabian Plate
Indian Plate
African Plate
North American Plate
Juan de Luca Plate
North American Plate
Caribbean Plate
Pacific Plate
Philippine Plate
Pacific Plate
Cocos Plate
South American Plate
Easter Plate
Nazca Plate
Australian Plate
Juan Fernandez Plate
Scotia Plate
Antarctic Plate

The Ring of Fire

Aleutia

Kurile Trench

Japan Trench

Izu Bonin Trench

Ryukyu Trench

Philippine Trench

Marianas Trench

● Challenger Deep

Bougainville Trench

Java (Sunda) Trench

Tonga Tren

Kermadec Tren

The Pacific Plate is one of the plates that make up the Earth's crust.

Did you know?

Most of the active volcanoes are actually under water!

ch

Puerto Rico Trench

Middle America Trench

ator

Peru - Chile Trench

There are more volcanoes along this border than anywhere else in the world. It's called the Ring of Fire.

DID YOU KNOW?

The word 'volcano' comes from the Roman fire god Vulcan.

500 million people live within the danger zone of a volcano.

Hot spots are places in the mantle where rocks melt to form magma.

The plates that make up the Earth's crust are called tectonic plates.

Visit a volcano

Many people want to see a volcano up close. But remember to never visit an active volcano because rocks and lava could fly up and hit you! If you visit a dormant volcano, always be careful and don't get too close.

Dangers of a volcano!

Ash falls are the most frequent volcanic hazard. Volcanic ash consists of pulverised rock, minerals and glass. It can damage engines, contaminate water supplies, and irritate our eyes, nose, and lungs.

Pyroclastic flows are clouds of super-hot gas and volcanic debris. They are extremely fast and the deadliest of all volcanic hazards. They occur during large eruptions.

Lava flows usually move slowly, so people have time to get away from them. But everything that stands in their way will get destroyed.

Lahars are volcanic mudflows. They are a thick mixture of mud, rock, and water. When it stops moving it solidifies — just like concrete! Lahars can be triggered without an eruption.

Volcanic gases can be toxic and deadly, but most of the volcanic gases are just water vapor! Sometimes toxic gases are trapped in low areas.

Landslides are masses of rock and soil that slide downwards because of gravity. Earthquakes, rainfalls, or volcanic activity can trigger a landslide.

The deadliest volcano

Mt Tambora in Indonesia has killed more people than any other volcano. After being dormant for centuries it erupted in 1815. Most people died of starvation. Ash clouds blocked the sunlight from reaching Earth, and crops wouldn't grow.

Massive eruptions can cause volcanic winters. Temperatures drop globally.

Did you know?

Volcanoes are active, dormant, or extinct. An active volcano has erupted recently. A dormant volcano has not erupted in a long time. An extinct volcano will never erupt again.

Extinct Volcano   Dormant Volcano   Active Volcano

Magma Chamber

Predicting an eruption

Scientists studying a volcano

There are often small earthquakes before an eruption. The ground around the volcano may swell. There are other signs, as well. The best way to foresee an eruption is to study the volcano's past behaviour. But we don't have detailed information about all volcanoes. This can make them difficult to predict.

Did you know?

All volcanoes are unique. They show different signs before erupting!

14

Evacuation

Sometimes, if it looks like a volcano is about to erupt, people will evacuate and leave their homes. Often people can return, but not always.

Not following evacuation orders can lead to death or serious injury!

Be prepared!

If you live near a volcano it's a good idea to speak with your family about what to do in case of an evacuation. Where will you go? How will you get there? What should you bring with you?

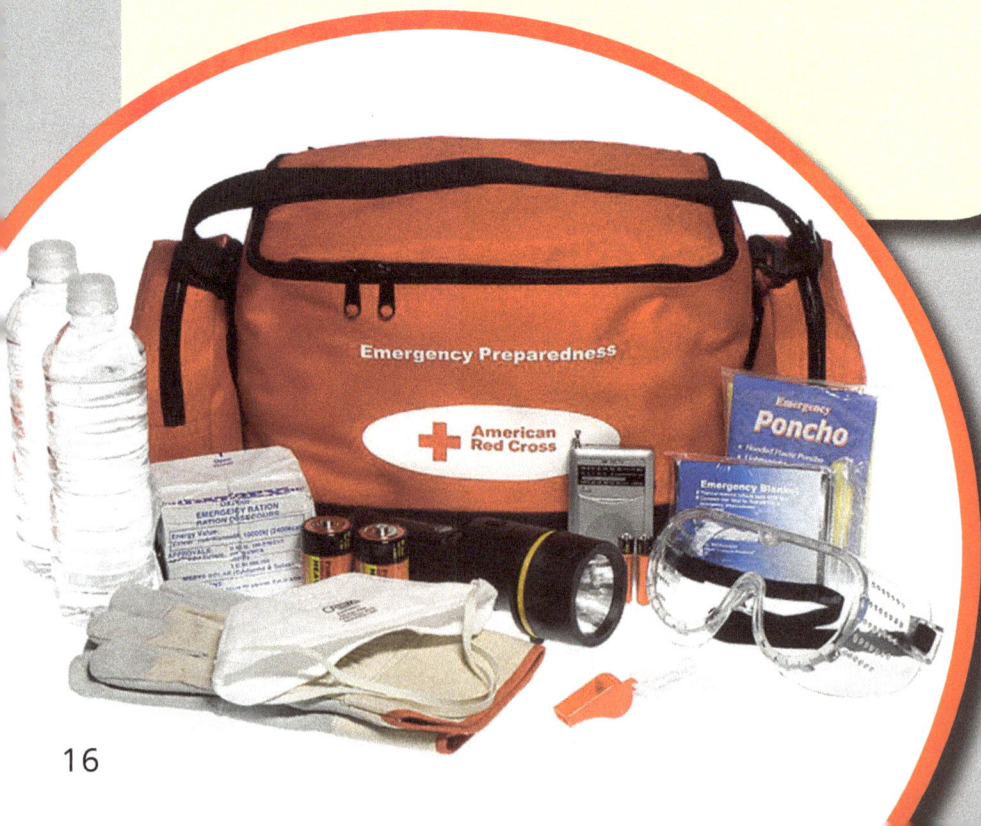

17

Help! The volcano is erupting!

1

Use goggles to protect your eyes.

2

Hold a damp cloth over your mouth and nose.

3

Keep as much of your skin covered as possible.

If you are outside:

1

Seek shelter.

2

Avoid low-lying regions.

3

Avoid downwind areas of the volcano.

If you are inside:

1

Close windows and doors, if you are not going to evacuate.

2

Stay inside until the ashes have settled, unless there is a danger of the roof collapsing.

3

Listen to the radio to know what is happening.

After the eruption

1

Help weak and injured people.

2

Avoid inhaling the ashes.

3

If evacuated, don't return until authorities say it's safe to do so.

It's Quiz Time

Q What is a volcano?

A An opening in the Earth's crust.

Q What are Earth's three layers?

A The crust, the mantle, and the core.

Q What is magma?

A Lava (molten liquid rock) inside the mantle.

Q Where do volcanoes form?

A In plate boundaries and hot spots.

Q What is volcanic ash made of?

A Pulverised rock, minerals, and glass.

Q What does 'evacuation' mean?

A To leave home because it's not safe anymore.

Q What places should you avoid during volcanic eruptions?

A Rivers and low-lying regions.

Q What should you do before a volcano erupts?

A Evacuate!

It's time to prepare for an emergency!

Use your finger to trace the lines. Which items should go into the emergency kit? Which ones should stay at home?

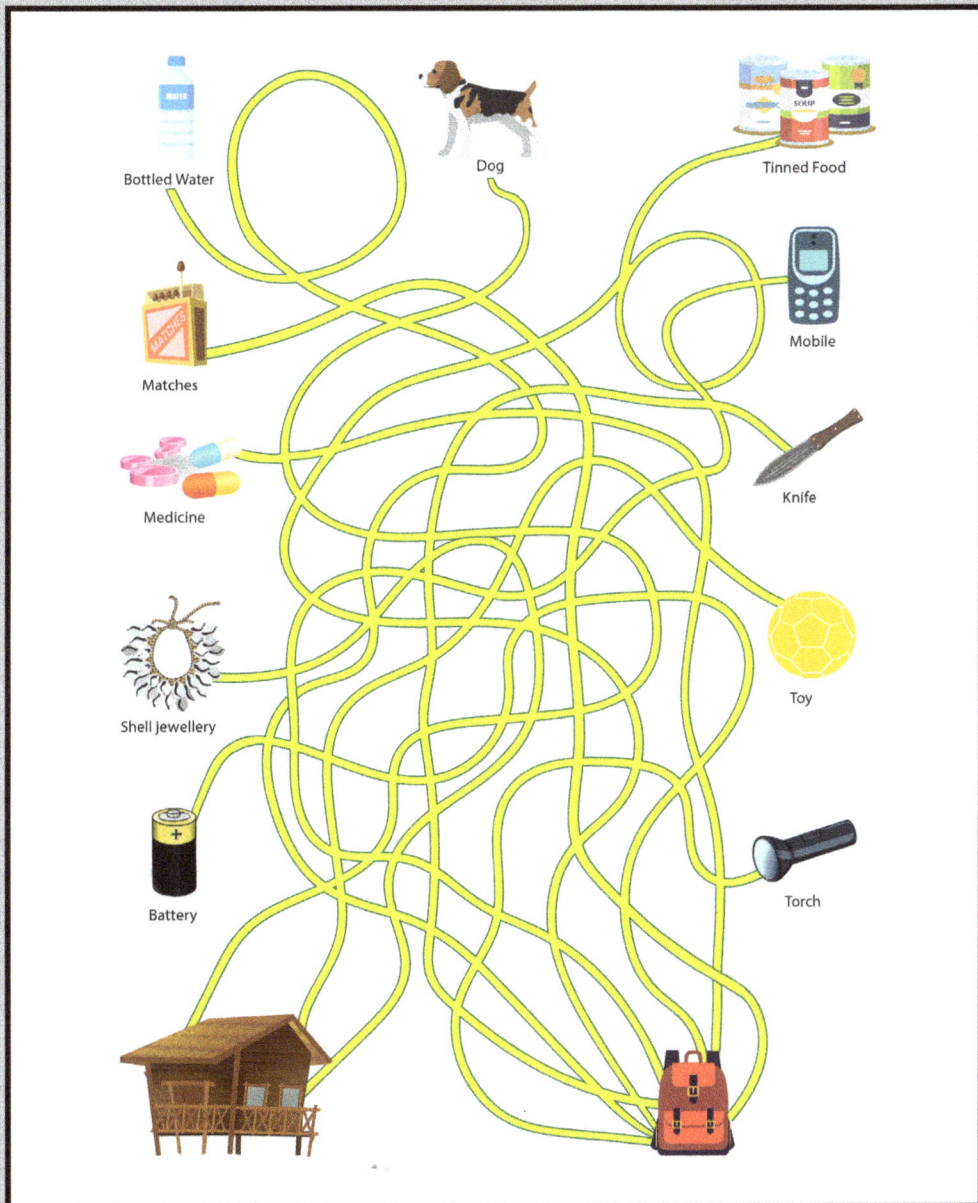

Bottled Water

Dog

Tinned Food

Matches

Mobile

Medicine

Knife

Shell jewellery

Toy

Battery

Torch

Emergency decision-making tree

Prior to the event of a tsunami, tropical cyclone, flooding, landslide or earthquake, speak with your family and teacher about your community's evacuation building or safe place.

Discuss how to respond to possible scenarios, and use the decision tree to help you decide the best course of action.

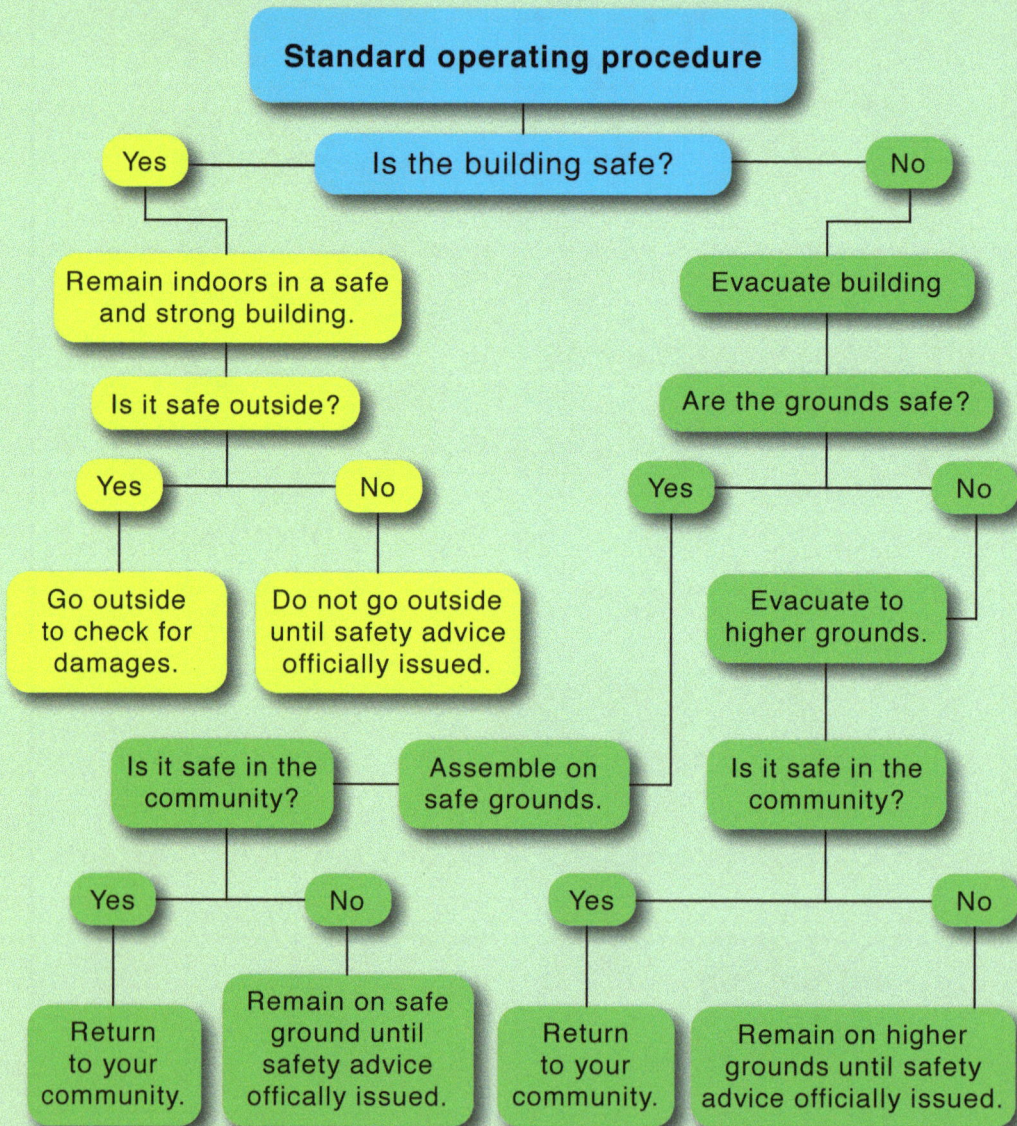

Standard operating procedure

Is the building safe?

Yes — Remain indoors in a safe and strong building.

Is it safe outside?

- Yes — Go outside to check for damages.
- No — Do not go outside until safety advice officially issued.

Is it safe in the community?

- Yes — Return to your community.
- No — Remain on safe ground until safety advice offically issued.

Assemble on safe grounds.

No — Evacuate building

Are the grounds safe?

- Yes — Evacuate to higher grounds.

Is it safe in the community?

- Yes — Return to your community.
- No — Remain on higher grounds until safety advice officially issued.

Supporting information

Emergency kit

Keep an emergency kit at home for your family.

The kit must contain:

First Aid Kit

Torch lamp

Radio

Batteries

Drinking water

Preserved food

Matches

Use the kit only in case of emergency and replace anything that has been used.

Shelter-in-place

Earthquake:
- Identify safe places where you can protect your head and avoid heavy falling objects.
- Don't forget an earthquake can cause a tsunami.
- If you feel a strong earthquake, go quickly to higher ground, and listen to the radio for warnings.

Tropical cyclone:
- Open louvers on the side of the building, away from wind to reduce the pull force of the wind on the roof.
- Remain calm, stay indoors but clear of doors and windows.
- Remain in the strongest part of the building.

Do not go outside until safety advice is officially issued.

Evacuate building

Assist people with disability and visitors.
Take your emergency kit.
Evacuate to higher ground and move to a safe location.

Tsunami:
- Run to a safe place in high ground or at least 2 km inside the island.
- Wait for at least 2–3 hours after the first wave to return to the village.

Listen to the radio for further information or reach out to the emergency contacts.

You can use these questions to talk about this book with your family, friends and teachers.

What did you learn from this book?

Describe this book in one word.
Funny? Scary? Colourful? Interesting?

How did this book make you feel when you finished reading it?

What was your favourite part of this book?

download our reader app
getlibraryforall.org

About the contributors

Library For All works with authors and illustrators from around the world to develop diverse, relevant, high quality stories for young readers. Visit libraryforall.org for the latest news on writers' workshop events, submission guidelines and other creative opportunities.

Did you enjoy this book?

We have hundreds more expertly curated original stories to choose from.

We work in partnership with authors, educators, cultural advisors, governments and NGOs to bring the joy of reading to children everywhere.

Did you know?

We create global impact in these fields by embracing the United Nations Sustainable Development Goals.

library forall.org

www.ingramcontent.com/pod-product-compliance
Lightning Source LLC
Chambersburg PA
CBHW040314050426
42452CB00018B/2839